LERNER PUBLICATIONS COMPANY
MINNEAPOLIS

Lerner Publications Company
A division of Lerner Publishing Group
241 First Avenue North
Minneapolis, MN 55401

Website address: www.lernerbooks.com

Library of Congress Cataloging-in-Publication Data

Domaine, Helena.
Robotics / by Helena Domaine.
p. cm. — (Cool science)
Includes index.
ISBN-13: 978-0-8225-2112-9 (lib. bdg. : alk. paper)
ISBN-10: 0-8225-2112-1 (lib. bdg. : alk. paper)
1. Robotics—Juvenile literature. I. Title. II. Series.
TJ211.2.H47 2006
629.8'92—dc22 2004013938

Manufactured in the United States of America
1 2 3 4 5 6 – BP – 11 10 09 08 07 06

Table of Contents

Introduction

Imagine, said Hollywood, that this is the future. Imagine a world where the robots have become as smart as the humans who created them. Imagine a rebellion where the robots turn against humans. When we watch movies such as *The Terminator* or *I, Robot,* we see this nightmare of robots running amok.

CAPTAIN FUTURE WIZARD OF SCIENCE

A comic book (*above*) and movie, *The Day the Earth Stood Still* (*above, right*), present typical 1950s images of creepy robots. In 2004, a police officer played by Will Smith (*below, right*) was still worrying about robots threatening humans in the movie *I, Robot.*

But we've also seen the future in a movie galaxy far, far away, where the robots were charming and loyal. In the Star Wars series, C3PO has a very human personality, and he is devoted to his human companions. R2D2 is adorable, but he is also brave and smart.

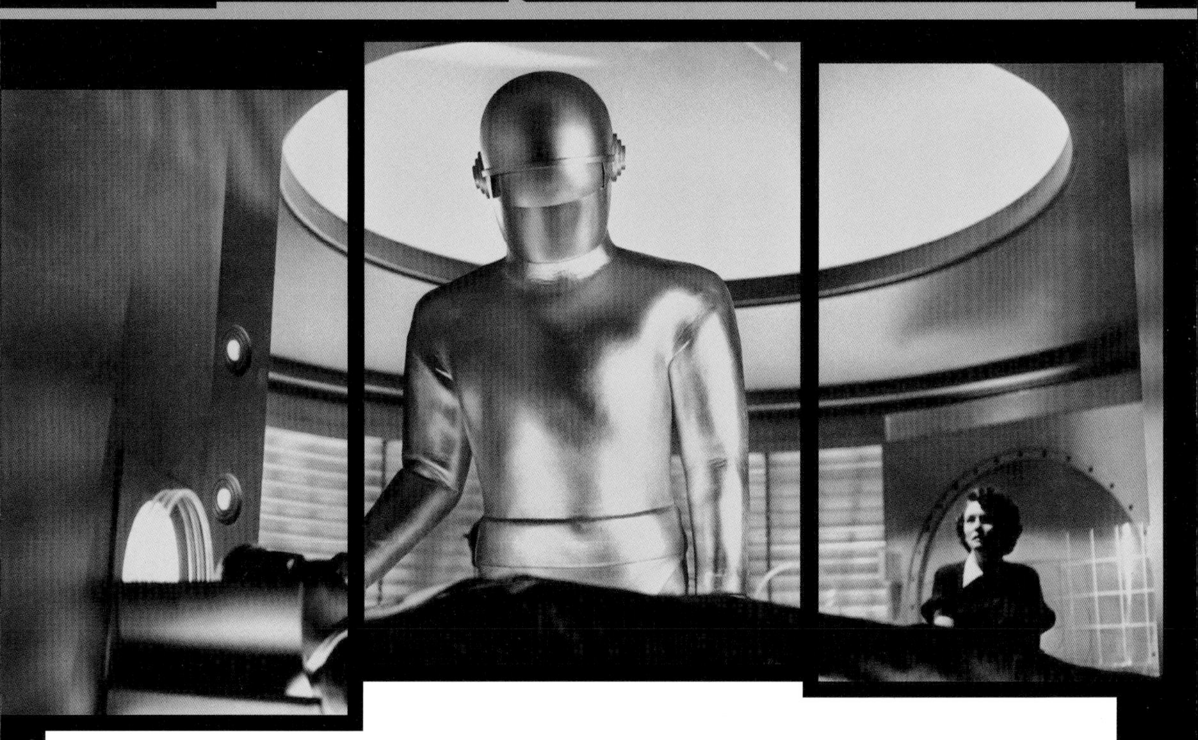

So which is it? Do we love robots, or are we afraid of them? Will the robots that scientists are developing and building chauffeur us around, clean up our messes, and save our lives? Or will robots somehow become a threat? Will they turn on us?

While Hollywood imagines a robotic future, science is working on a robotic present. Roboticists—scientists who specialize in the creation of robots—have quite an imagination too. They imagine and create robots designed to help humans in everything from office work to space exploration. And they imagine ways to make robots likable and fun to have around. You may find that what goes on in an average robotics lab is better than any Hollywood movie!

chapter 1

How Robotics Began

Humans have always dreamed of building an artificial person, a completely selfless machine that looks like us and does our bidding. Ancient legends tell of people making little creatures called homunculi out of clay, sand, or dust. The homunculi could be brought to life with magical spells and sacred words.

In the first century B.C., Greeks created small devices that could move on their own. But the Greeks did it with science, not magic. They used their discoveries of pneumatics and hydraulics—using air and fluids to operate machines. The Greeks called their creations automata, meaning "self-activating." The Greek automata were certainly clever devices. But the Greeks considered the automata works of art. The devices were not yet robots.

In the 1400s A.D., Italian artist and inventor Leonardo da Vinci pictured a much more advanced machine. His notebooks contain detailed

FUN FACT!

A robot is defined as an automatic machine that can be programmed to perform certain tasks or jobs. Many robots have humanoid characteristics, meaning they resemble humans. They may have humanlike faces or limbs. Androids are robots built to have human bodies. They move and function as humans.

sketches of a robot knight run by mechanical disks, gears, and pulleys. But da Vinci never built his robot knight. Five hundred years later, an engineer from the United States, Mark Rosheim, studied da Vinci's plans at the Institute and Museum of the History of Science in Florence, Italy. From a modern engineer's viewpoint, da Vinci's design made sense. In 2003, Rosheim and a team of engineers built the robot knight for a British television series. Put to the test, the robot knight stood, sat, and raised its arms. The test proved that da Vinci was the first person to design what can technically be called a robot.

Fast Thinking

For centuries after da Vinci, people continued to develop methods of mechanization—building machines that ran automatically, with little help from human workers. Inventors worked to create devices, from wristwatches to adding machines to weaving looms, that were efficient, reliable, and capable of running on their own. Mathematicians, too, worked on new systems for making complex calculations (math and logic problems). These developments in technology and math led to a new field in the 1900s: computer science.

Computer science went beyond mechanization to study how information itself could be automated. Computer scientists studied how

The Scribe

Swiss clockmakers and watchmakers have long been masters of miniaturization. In 1772, the Swiss inventor of the wristwatch, Pierre Jaquet-Droz, combined miniaturization with automation to create a robotic doll called the Scribe. The Scribe mimics the act of writing. Its head and eyes follow the motion of its right hand as it dips a pen into an inkwell. The Scribe then carefully writes words letter by letter. A disk inside the Scribe's body determines which word the Scribe writes. The disk has 40 interchangeable pegs, and the pegs can be set to reproduce the characters of any language. In other words, the Scribe is programmable, just like a computer—except, of course, that the Scribe was invented almost two centuries before computers.

The Scribe (left) still works and is on exhibit at the Museum of Art and History in Neuchâtel, Switzerland. The exhibit also includes Jaquet-Droz's other mechanical figures: the Musician, the Artist, and the Draughtsman.

machines could be designed to make calculations automatically. They then studied how data (information) could be electronically fed into and stored in the computer. With that capability, the computer could apply its calculations to any number of problems. Think of it this way: A clock is mechanized. You set the time, and then the clock's parts work together to keep track of it. A computer can be programmed to tell you what time it is too. But it can also be programmed to calculate what time it is in China, what time it will be after 1,564 seconds have gone by, or exactly what time the sun will rise and set on December 17 of next year.

Very early computers made their calculations accurately but not very quickly. The computers were also very large. Some took up entire rooms. They were filled with vacuum tubes, electric switches, and wires. They used piles of punch cards, spools of paper, and reels of magnetic tape. And each computer had to be programmed step-by-step.

As computer science developed, it became focused on making computers faster, smaller, and easier to use. Researchers worked on miniaturizing

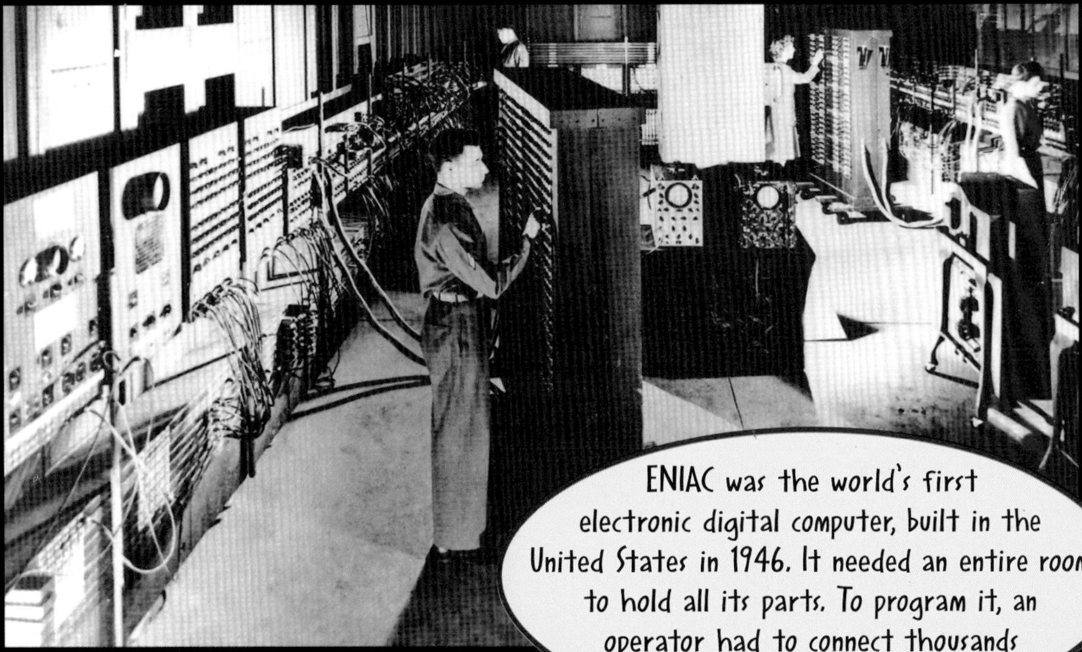

ENIAC was the world's first electronic digital computer, built in the United States in 1946. It needed an entire room to hold all its parts. To program it, an operator had to connect thousands of wires by hand.

(reducing the size of) computer parts. All the tubes, switches, and wires were replaced by tiny transistors and electric circuits stored in a small piece of material called a computer chip. Computer chips greatly increased computer speed and memory (storage room for programs and data). Miniaturization took computers from the research laboratory into offices, banks, and factories. But miniaturization also paved the way for computers to be used in robotics. Once computers were small enough to fit inside a robot, a whole new world of possibilities opened up.

Working for a Living

The word *robot* comes from the Czech *robata,* meaning "forced labor." The word was first used in 1920 by Czech playwright Karel Capek in *R.U.R. (Rossum's Universal Robots).* In Kapek's play, the Rossum Company manufactures robots to do human work. And actually, robots have been used in industry and manufacturing for decades. In fact, almost everything we eat, wear, or use was worked on, at some stage, by a robot.

In 1913, Henry Ford started using assembly lines in his car factory near Detroit, Michigan. On an assembly line, a car moves down a long line of

A robotic arm in a factory processes silicon wafers for use in industry.

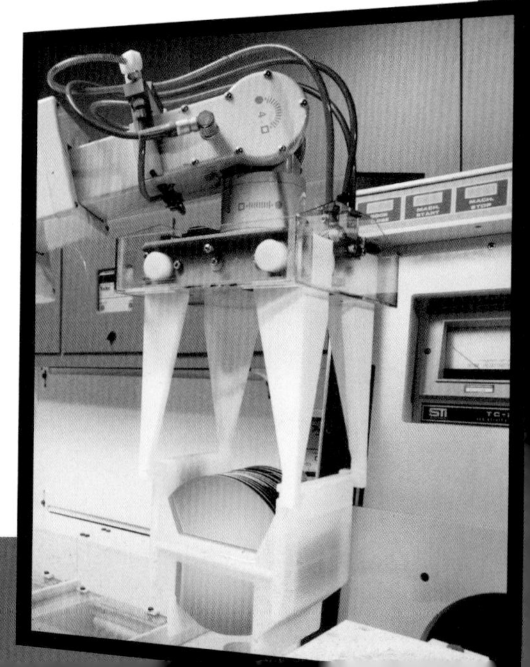

workers who each perform a specific task until the car is complete. Ford loved the efficiency and precision of the assembly line. So he would have loved robots, which make up 50 percent of the modern automobile industry's labor force. Robots do very well at assembly-line work.

Robotic manipulators weld cars along an automated assembly line.

Most assembly-line robots are computer-operated, multi-jointed arms called manipulators. They do repetitive and dangerous jobs along the assembly line, such as spot welding car bodies, spraying paint on chassis, and pouring rubber into tire molds. But robot manipulators also work in paper mills, bottling plants, toy companies, and factories that make computer parts. These manipulators lift, bolt, fold, stamp, package, weld, remove, and replace, hour after hour, for two or three shifts a day.

The manipulators are designed to mimic the motions of the human arm. Some have a ball-and-socket assembly, much like the human elbow. Other manipulators lengthen and pull back by telescoping—that is, their cylindrical sections slide over one another. A third type of robotic arm looks like a huge snake. Robots outfitted with four or six of these arms look like metallic aliens.

The arms end in grippers—crude versions of human hands. To be effective, the grippers must match the shape of the item they have to pick up, use, or manipulate. Most industrial robots do fine with their grippers, but scientists are developing much more humanlike robotic hands. These hands have four fingers and a thumb—capable of providing a grip just like a person's hand.

Flex Those Muscles!

Professor Yoky Matsuoka of Carnegie Mellon University in Pittsburgh, Pennsylvania, and her team of graduate students are building an anatomically correct testbed (ACT) hand. The ACT hand uses mechanical devices called actuators to work like human muscles and tendons. Matsuoka hopes that the ACT hand can be studied to understand all the complex movements of the human hand. Surgeons may also use the ACT hand as a model for reconstructing human hands damaged by injury or disease.

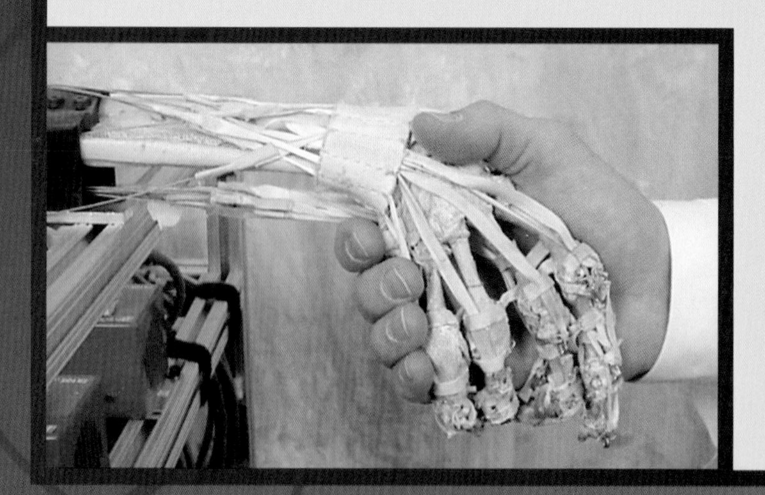

The ACT hand

The DARPA Challenge

The auto industry may have been quick to use the first robots. But it was a group of students, computer scientists, and robot builders who met a 2004 challenge to build a robotic car. To encourage research on unmanned vehicles, the U.S. Department of Defense (DOD) set up a race in California's Mojave Desert. The DOD's Defense Advanced Research Projects Agency (DARPA) offered a $1 million prize to the builders of a robotic vehicle that could find its way, unaided, across 142 miles (229 kilometers) of the Mojave's rough terrain. Thousands of spectators lined the course on the morning of March 13, 2004. The robo-vehicles left the starting line packed with computers, sensors, and elaborate vision systems. Within a few hours, all the vehicles had either crashed or lost their way. So there was no $1 million winner that day. But competitions and exhibitions like the DARPA Challenge help researchers test new robotics technology.

A robotic car heads to the DARPA starting line.

Roboticists continue to work on creating robots or robot parts that are specially designed for certain types of work. As with the manipulator's "hand," roboticists often use human or animal models to understand motion, balance, and flexibility. The designs roboticists have come up with have taken robots beyond the factory into some pretty amazing places.

chapter 1

Working Robots

There are lots of places we'd like to go but can't. Dangerous places. Distant places. Inaccessible places. We can explore these places by sending in robots. These mechanical daredevils and adventurers have computer brains that don't feel fear or panic. Killer levels of radioactivity? No problem. The black, airless vacuum of space? The crushing pressure of tons of ocean water? Tiny passageways through ancient rock? Bring it on, say these brave new robots.

Andros 5, for example, handles live bombs for the Baltimore (Maryland) Police Department. Rosie, built by a team at Carnegie Mellon University in Pennsylvania, can safely roll into highly contaminated nuclear facilities and wash them down or take them apart. Houdini, who might be considered Rosie's baby brother, can enter hazardous waste storage tanks to tackle cleanups.

Robots have also been used in war zones to find and deactivate land

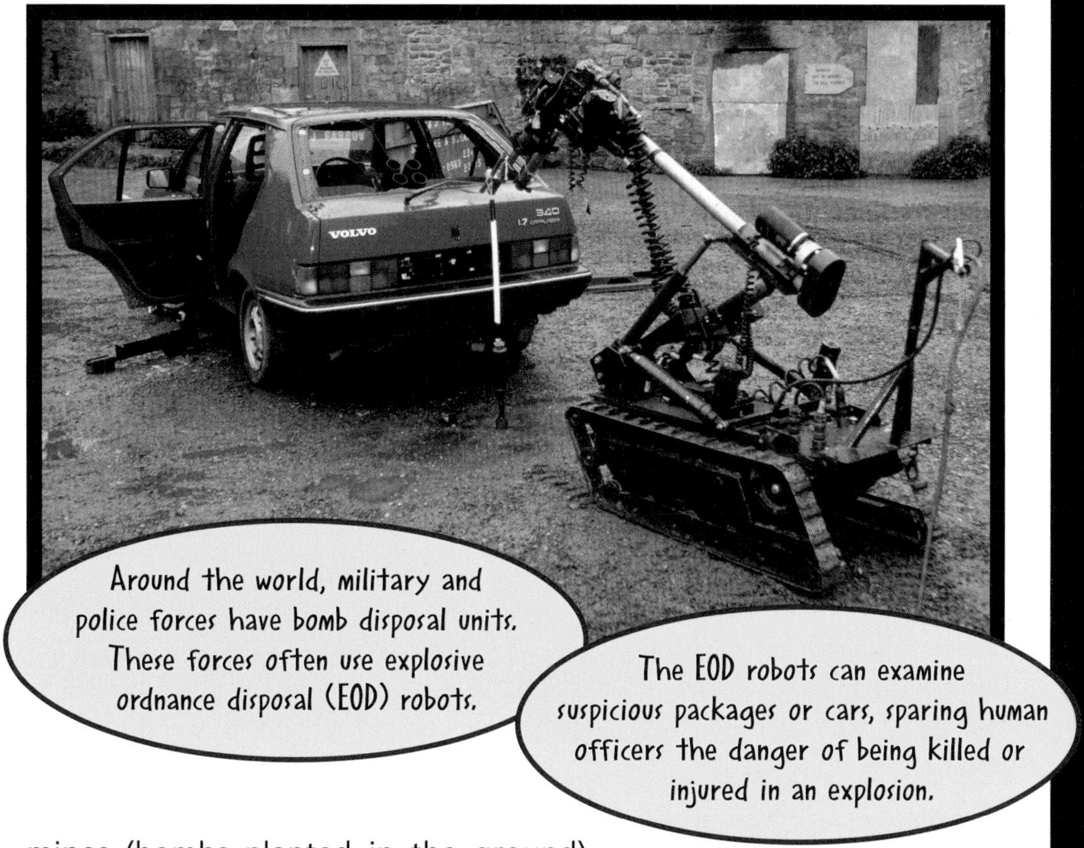

Around the world, military and police forces have bomb disposal units. These forces often use explosive ordnance disposal (EOD) robots.

The EOD robots can examine suspicious packages or cars, sparing human officers the danger of being killed or injured in an explosion.

mines (bombs planted in the ground).

Robots crept through the choking rubble of the World Trade Center towers after the September 2001 terrorist attack in New York City. Robots have worked at nuclear accident sites in air so contaminated that the robots couldn't be returned to their labs. Robots have been to Mars, and soon they will crawl through human arteries to break up life-threatening blood clots.

You Want Me to Go Where?

In 1994, the National Aeronautics and Space Administration (NASA) teamed up with scientists at Carnegie Mellon University and the Alaska Volcano Observatory to send a robot to explore an active volcano. Scientists explore volcanoes to learn how they work and how to read the warning signs of a volcanic eruption. An eight-legged robo-walker named Dante II climbed down the throat of Alaska's Mount Spurr, 90 miles (145 km) west of Anchorage. Dante's mission was to explore the crater floor and take gas and soil samples. It was something that no human had any intention of doing—not even for the good of science.

Dante II makes its way across a snowy path in Alaska.

Dante's designers knew the descent would be very tricky. The north wall of the volcano has a perilous 1,000-foot (305-meter) drop, and the south wall is steep and covered with rocks. The designers gave Dante a system of servomotors, mechanisms that help Dante's main computer. The servomotors allow Dante to raise and lower each leg individually as the robot climbs over rocky surfaces. Dante's footpads and legs also have sensors to keep it from crashing into rocks or falling into holes.

But even with all this high-tech robo-anatomy, nobody trusted Dante to make its own decisions. Dante was connected to its human team by satellite and the Internet. Its main computer analyzed every step before

Why Study Volcanoes?

Volcanic eruptions can kill people and destroy property. The most violent eruptions can erase entire cities. They can cause huge ocean waves, floods, and mud slides. Volcanoes can also disrupt weather patterns around the world. Volcanologists (scientists who study volcanoes) can't control eruptions. But through research, volcanologists have learned how to read a volcano's warning signs. Communities near active volcanoes can come up with emergency plans for alerting and evacuating people. In providing research aid to volcanologists, robots help people plan for these natural disasters.

Mount St. Helens, an active volcano south of Seattle, Washington, belches steam in October 2004.

it allowed the robot to go forward. This made Dante one poky robot, but it eventually reached the floor of the crater, safe and sound.

As Dante gathered samples, the robot's stereo cameras relayed a realistic three-dimensional view to the computer screens in front of the scientists at the volcano's rim. And thanks to something called Virtual Environment Vehicle Interface software, the scientists felt as if they were right there in the volcano with Dante.

But a near-perfect robotic adventure ended in a way familiar to anyone who's ever climbed a steep hill. Dante slipped in some loose dirt on the way out of the volcano and could not regain a foothold. A short time later, the science team called in a helicopter to rescue the stranded robot.

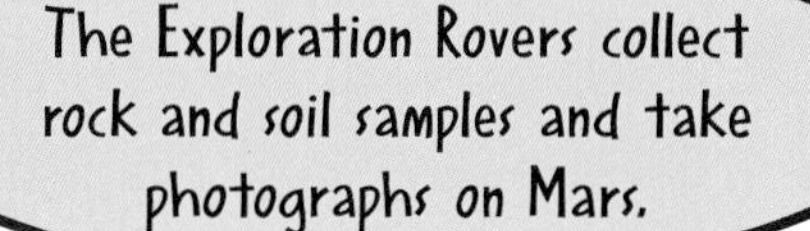

The Exploration Rovers collect rock and soil samples and take photographs on Mars.

Unfortunately, no one can chopper to Mars to save the Exploration Rovers if these robots get into trouble. NASA landed twin Rover robots, Spirit and Opportunity, on Mars in 2004 to explore the planet, collect soil and rock samples, and take photographs. Spirit and Opportunity are all alone on the red planet, millions of miles from Earth. And Mars is a far more hostile environment than the inside of a volcano. Mars is very cold, averaging –67°F (–55°C). Its strong winds whip red dust across the rocky surface of the planet.

The Rovers' connection to NASA is tricky too. Communication between the robots and NASA scientists are sent through millions of miles of space via computer connections to orbiting spacecraft and antennas on Earth. As the Rovers roll across the Martian landscape, any helpful messages from their human teammates on Earth are delayed by several minutes. So the Rovers are designed to make many of their own decisions. They are given assignments, but it is up to them to figure out

Phoning Home

The distance from Mars to Earth varies from 110 to 200 million miles (177 to 332 million km), depending on the planets' orbits. How does NASA maintain communications with the Mars Rovers over that distance? Communications links are provided by NASA's Deep Space Network (DSN). DSN is an international system of very large and powerful radio antennas. The network's communications centers are located in California's Mojave Desert; near Madrid, Spain; and near Canberra, Australia. The communications centers can send commands directly to the Rovers, and the Rovers can send data directly back. But often the communications centers relay messages between NASA and the Rovers through other spacecraft orbiting Mars. Relaying messages through Mars orbiters takes less energy for the Rovers.

DSN antennas range in size from about 110 feet (34 m) to about 225 feet (67 m) across. They can send signals to and pick signals up from millions of miles away. The Mars Rovers also have four small antennas on board. They can send four different signals to four places at the same time.

how to get them done. The Rovers also have a "survival instinct" programmed into them, helping them adapt to unexpected situations, such as changes in the terrain.

FUN FACT!

In the robot world, there are three degrees of independence. A teleoperated robot relies on humans to make all the decisions, but a computer helps the robot carry them out. With a supervisory robot, humans make the key decisions, but the robot can work without constant guidance. An autonomous robot has total control over its actions.

A Tight Fit

Robots have helped us explore live volcanoes and hostile planets. There are other places to explore that are not dangerous but are simply out of reach to humans. One of those places is inside the Great Pyramid of Giza in Egypt. Modern explorers have long been fascinated by the pyramids—huge structures built thousands of years ago as tombs for Egyptian rulers. Even after hundreds of years of excavation and study by archaeologists, the pyramids still have their secrets.

The Great Pyramid was built around 2500 B.C. for King Khufu (also known as Cheops). Inside the pyramid, four narrow

The Great Pyramid is the largest pyramid ever built. At its peak, it is almost fifty stories high.

shafts lead out of two rooms known as the King's and the Queen's Chambers. No other pyramids in the region contain shafts such as these. The shafts are about 9 inches (23 centimeters) high and 8.5 inches (22 cm) wide. They rise at steep angles from the chamber walls. The shafts in the King's Chamber open to the outer walls of the pyramid. But the shafts in the Queen's Chamber end somewhere inside the pyramid.

When German engineer Rudolf Gantenbrink first visited the pyramid in 1991, he was struck by the mystery of the shafts. It must have been very difficult for the pyramid's builders to include the shafts in their design. So Gantenbrink believed they must have been important. But what was their purpose? And how could anyone explore something so deep and narrow? Gantenbrink thought about it. When he returned to Germany, he began work on a robot small enough to fit in the pyramid's shafts.

After some trial runs, Gantenbrink developed Upuaut-2. Upuaut-2 is made from aircraft aluminum and weighs only 13 pounds (6 kilograms). It is equipped with headlights, a laser guidance system, and a miniature video camera. Seven electric motors drive the wheel/track systems, and power is supplied through a cord connected to a generator. Upuaut-2 is also connected to a computer with a video monitor.

Upuaut-2 was only the first pyramid explorer. Other researchers have since sent similar robots into the shafts.

In March 1993, Gantenbrink sent Upuaut-2 into the southern shaft of the Queen's Chamber. Gantenbrink hoped that the videos Upuaut-2 took would solve the mystery of where the shafts ended.

But as it turned out, the little robot's journey only deepened the mystery. As Gantenbrink watched outside on the video monitor, Upuaut-2 climbed about 200 feet (61 m) up the shaft. Suddenly, Upuaut-2's camera showed that the robot had been stopped by a stone block in the shaft. The stone was smooth and had copper decorations. It had obviously been placed there on purpose, high up inside the narrow shaft. But by whom, and why? Until another robot with different equipment explores the shaft, the mystery remains.

The Incredible Shrinking Bot

Upuaut-2 was certainly in a tight spot, but an 8-inch (20-cm) shaft is a roomy cruise compared to scientists' plans for the snake-bot. Scientists at the California Institute of Technology are working on the designs for a tiny snake-bot to travel through the human gastrointestinal system (the stomach and intestines). As a doctor looks down a patient's throat for swelling or other signs of illness, the snake-bot would look at a patient's insides. A camera and sensors would help the snake-bot gather medical information for doctors. The snake-bot's information would help doctors diagnose disease and may even help in therapy.

Designed for rescue operations, large snake robots (*below*) can, for example, slither through earthquake rubble to search for survivors. Miniature robots (*above*) from New Mexico's Sandia Laboratory also explore tight spots.

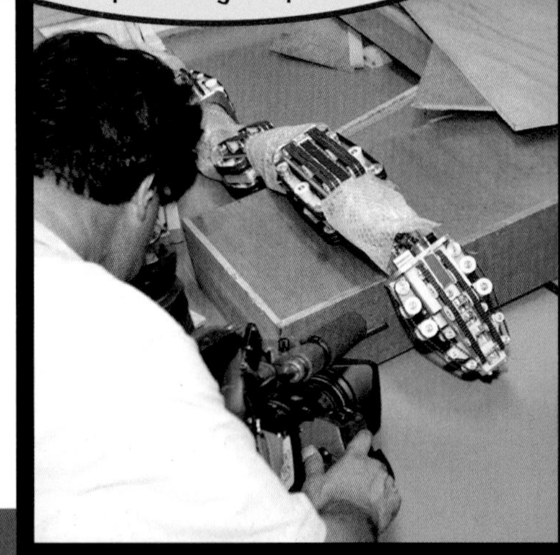

But without question, the tiniest and most daring medical robots are being designed in Sweden. The Swedish micro-bots are smaller than the hypen between *micro* and *bots* in this sentence. The micro-bots are made of silicon coated in gold and then encased in polymer (a plastic compound) that can be made to shrink or swell. This allows the segments of the robot to bend so it can pick things up and move them around.

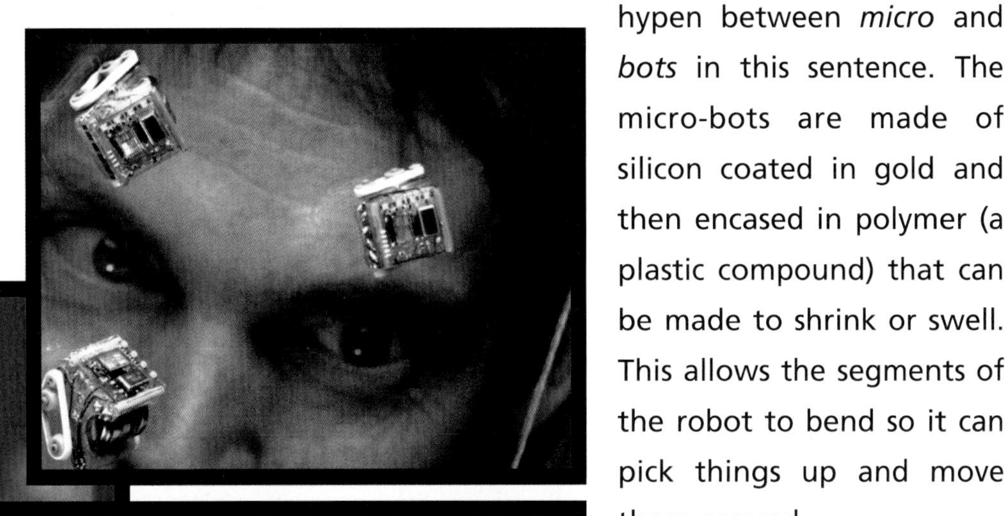

Sandia researcher Doug Adkins (above) designed the miniature robots to work in swarms, like insects. They communicate with each other and with a central station.

The Swedish micro-bots are designed to be submersible, which means that they can operate in all kinds of fluids. The research team imagines a time in the not-too-distant future when the micro-bots can be injected into the human bloodstream. Doctors hope the micro-bots will be able to clean up the plaque that causes heart attacks and break through the blood clots that cause strokes. The micro-bots could also remove bacteria and maybe one day even fix disease-causing cells.

In the old sci-fi movie *Fantastic Voyage,* five scientists and their submarine, the *Blood Vessel,* were shrunk to microscopic size and injected into the bloodstream of a fellow scientist. Their mission was to reach an inoperable blood clot in their friend's brain and save his life. What Hollywood imagined as movie fantasy in 1966 is becoming science fact.

chapter 3

Improving Robots

A sign on the chain-link fence surrounding a factory reads, "This area patrolled by security guards." But the guards are not what you'd expect. Instead of uniformed, gun-toting humans, the security force is made up of wheeled robots. The robots use sonar—navigation by sound—to make their way through the empty plant. If one of these robo-guards comes across an intruder, it can alert the police. But it won't be able to wrestle the burglar to the ground or even chase him. Wheeled robots are very slow.

In a hospital, several robot cafeteria workers deliver meals to the patients. They roll from room to room, neatly avoiding objects and staff members. When necessary, they can even ride the elevator. But if something falls off their food cart, they can't pick it up. They can't help patients open the cellophane packet containing their plastic utensils. And in one case, a service robot, unable to stop itself, pitched head over wheels down a stairwell. Wheeled robots can be very klutzy.

These are just two examples of service robots designed to provide daily help to humans. But roboticists know that service robots have a long way to go. The robots will need to be able to negotiate their way around without tripping or becoming confused. They'll need problem-solving skills. Robotic nurses will need legs and arms. They'll have to be able to tell the difference between a patient who is flushed with fever and one who is as pale as a ghost, and decide if one, both, or neither needs help. So, in dozens of laboratories in the United States, Canada, Japan, and Australia, scientists and engineers are trying to design robots based on the best models of all—those found in nature.

In the early 2000s, some hospitals in China began using robots to perform tasks and carry information.

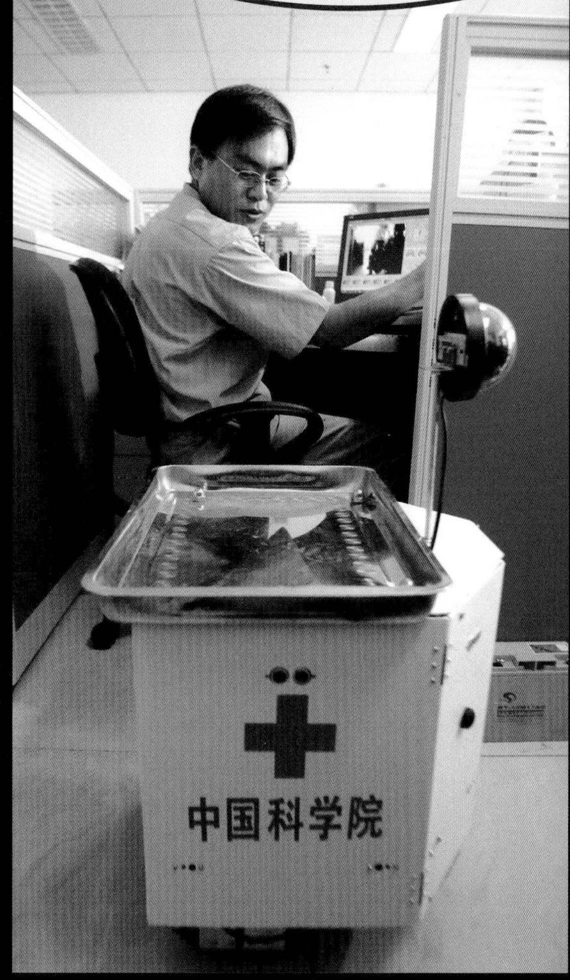

First Steps

At the University of California, Berkeley, biologist Robert Full is in cockroach love. For several years, he has been studying this less-than-adorable creature to find out how it is able to move so fast. After videotaping cockroaches working out on little treadmills, Full discovered that a cockroach doesn't actually run. It bounces, like a pogo stick. This sounds like an unsafe way to travel, but the cockroach solves the problem by using a tripod ("three feet") gait. It has six legs—all insects do—but only three are actually on the ground at any one time. In terms of balance, a tripod is very stable.

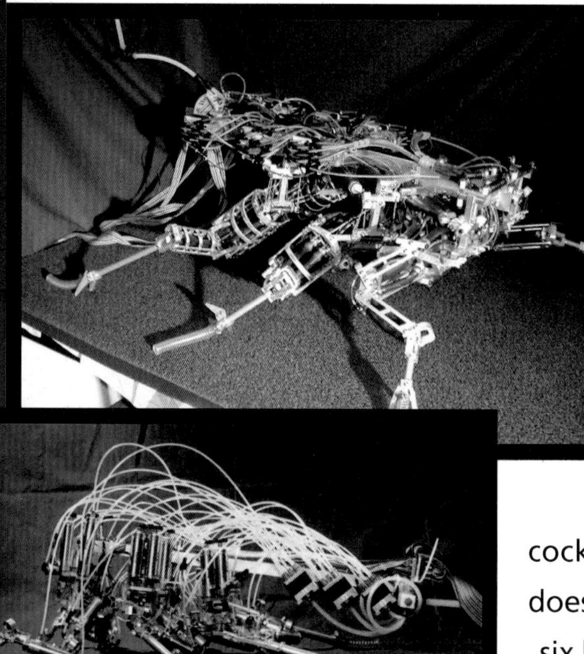

Engineers at Case Western Reserve University (CWRU) in Ohio built a robotic cockroach. The Case robo-roach doesn't have a head, but it does have six legs and a platform body to carry the electronics. It has a tripod gait, and it bounces, just like a real live cockroach.

At CWRU, Daniel Kingsley, Roger Quinn, and Roy Ritzmann studied real cockroaches (*below*) to develop their robotic version (*above*).

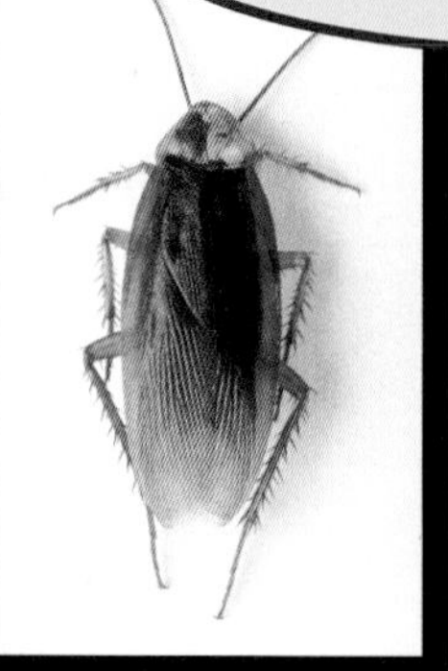

Scientists have built several different six-legged creatures, called hexapods. Without a doubt, one of the most interesting is Ariel, the robot crab. Ariel was developed by iRobot to scuttle along the beach and on the ocean bottom, searching for underwater explosives.

Ariel walks sideways, just like a real crab, and can scramble over uneven ground. Pebbles and ocean surf are not a problem. But the niftiest part of Ariel is the way it can adjust to life's little accidents. If Ariel loses a leg, it can quickly adapt to a five-legged walk. And if Ariel is upended, instead of lying helplessly on its back, its legs flip down. The top becomes the bottom, and Ariel keeps on going.

You may be thinking that if scientists can build robots with six legs, a two-legged (bipod) robot ought to be easy. But when it comes to legs, the fewer there are, the harder it is to keep the robot stable. Unlike the cockroach, which always has three legs on the ground when it is moving, a bipod has only one. So with every step, a two-legged robot must somehow reestablish its balance, or it will fall over.

Harder than It Looks

Nobody teaches us how to walk. Walking is something we more or less figure out on our own when we're about a year old. After our leg and back muscles have grown strong enough to support our weight, we pull or push ourselves to a standing position. Then comes the hard part. We want to move forward—to where Mom or the family dog or that shiny toy is. But making it happen takes a little practice. That's because we're bipeds. When we walk, we raise one leg to step forward. All our weight has to be supported by the leg on the ground. But then when we take a second step, we change feet, and all our weight has to quickly shift. Learning to shift our weight back and forth is what keeps us from toppling over. After we learn to walk, we branch out to running, jumping, hopping, and other motions. Those all involve slightly different skills in muscle control and balance.

Just like us, bipedal robots have to redistribute their weight as they move. But they don't have our learning instincts or sense of balance. To get bipedal robots moving, roboticists study how our muscles and joints work, and what techniques we use for balance and control. All this information has to be translated into robot anatomy and computer programs.

Balance is a tough thing to program into robots. But researchers at the Massachusetts Institute of Technology (MIT) Leg Laboratory are making remarkable progress. They have built a lot of weird and wacky robots. Experimenting with these walking, running, and hopping robots helps researchers find the most stable and practical leg designs.

FUN FACT!

The study of how to create robots that can mimic the movements of animals, fish, or insects is called biomimetics.

Monopod is possibly the strangest robot because it is just one leg and one weight to provide balance. But Monopod has joints and a springy foot similar to what humans have. It paved the way for Spring Flamingo, the first two-legged lab robot to use its ankles and feet as part of the dynamics of walking. 3D Biped can perform somersaults, and Uniroo imitates the bounding gait of a kangaroo.

An MIT roboticist takes a test walk with Troody the dinosaur robot. Troody was modelled after a real dinosaur called a *Troodon*. Researchers at MIT's Leg Lab studied *Troodon* skeletons found by archaeologists, to understand how *Troodon* moved.

But one of the most exciting robots at the MIT lab is Troody, a two-legged dinosaur robot. Troody is so lifelike, it could hitch a ride back to the Cretaceous Era and fit right in.

Together, the MIT Leg Lab robots have strolled, hopped, trotted, bounded, and actually run along simple paths, up stairs, and over obstacles. One robot hit a running speed of 13 miles (21 km) per hour. Programming a robot to walk like a real animal may be difficult, but it is far from impossible.

ASIMO

In October 2000, Honda Motor Co., Ltd., introduced ASIMO (which stands for Advanced Step in Innovative Mobility)—the world's most advanced bipedal humanoid robot. ASIMO is exactly 4 feet (1.2 m) tall and weighs 115 pounds (52 kilograms). It looks like a spaceman, complete with backpack. (That's where its computer is located.) Honda engineers spent nearly fifteen years of study, research, and development to create ASIMO. They would probably say that not a minute has been wasted.

When Honda engineers set out to create a bipedal humanoid robot, they used motion technology to analyze the way human beings walk. A key question was, how do humans transfer weight from one leg to the other when we walk? The engineers worked on ASIMO's legs, which had to be able to make the same complex movements as human legs. Then, as each body part was added to create a humanoid robot, the engineers made adjustments for the new structure and additional weight.

Just like us, ASIMO controls its own walking, and it can stroll along at a leisurely one mile (1.6 km) an hour. When ASIMO made its debut at the

Frankfurt Motor Show in Germany on September 9, 2003, the robot proudly showed off all its skills. It walked backward, turned corners, climbed steps, shook hands with people, and even recognized the faces of a few old friends, addressing them by name. And for the big finish, it even danced around a bit. In 2004 ASIMO was upgraded with a rotating hip that allows it to jog at 1.9 miles (3.1 km) an hour.

ASIMO's amazing accomplishments are the result of the latest developments in electronics, computer science, voice and sound recognition, and motion technology. ASIMO can even use networks, such as the Internet, to help it answer people's questions. But what so many visitors to the show found truly remarkable was ASIMO's confident and nearly human stride, a major breakthrough in robotics.

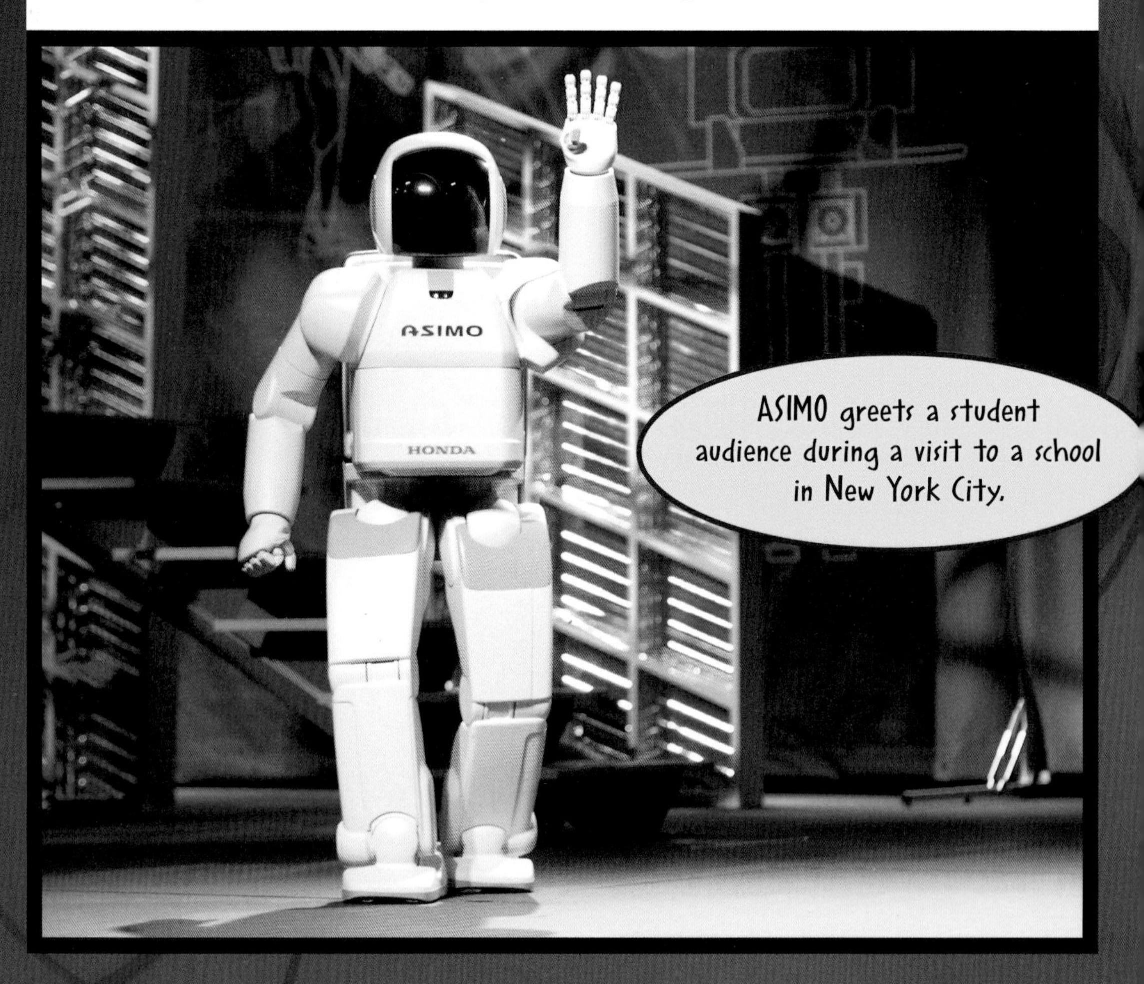

ASIMO greets a student audience during a visit to a school in New York City.

Artsy Robots

The Sony Corporation's QRIO robot took center stage—literally—in March 2004, when it conducted the Tokyo (Japan) Philharmonic Orchestra. QRIO can perform many tasks. But Sony, a Japanese electronics company, wanted to show off the robot's ability to control its motions. QRIO held a conductor's baton and led the human musicians through Beethoven's Symphony No. 5. Japanese automaker Toyota has also proudly produced a musical robot. The Toyota robot can play "When You Wish Upon a Star" on a trumpet. Toyota says it hopes to soon have an entire robot band ready to belt out tunes.

QRIO (right) and Toyota's trumpet player (below) were both designed to test controlled robotic movement.

Who's Got the Ball?

Robots aren't all work and no play. On May 4, 2003, robots from around the world played soccer in the International RoboCup Federation's American Open, held at Carnegie Mellon University. Hiroaki Kitano had established RoboCup in 1997 with the hope that it would lead to the development of robotic soccer players good enough to play against—and beat—human athletes.

That first 1997 tournament was a little chaotic. The robots had a tough time finding the ball, recognizing their teammates, and figuring out which goal they were supposed to aim for. But as engineers improved the robots' vision systems, play improved. By the 2001 games, the 8-inch-tall (20-cm), wheeled robots in the Small League were playing two ten-minute halves on a field the size of a Ping-Pong table. (Their soccer ball was an orange golf ball.)

Aibos, robotic dogs, compete in a RoboCup soccer game.

The Sony Corporation sends its Aibo team to the Open. Most RoboCup players are of the two-legged variety, but the Aibos are little robotic dogs. The Aibos whack the ball by getting down on their elbows. This tail-in-the-air position gives them enough stability to use both front paws. Play is slow and a bit goofy, but the Aibos are, after all, still amateurs.

Aibos have many motors that allow them to walk, stand, sit, and lie down. Aibos are also programmed to bark, find and play with their toys, and take "naps" when their batteries need recharging. And like real dogs, they will try to get attention from their owners.

But as wonderful as ASIMO is, it is still, well, robotic. It looks and acts like a robot. And ASIMO lacks the one thing that humans most respond to in each other: a face. What is going on, we might wonder, behind the black, featureless plate that hides ASIMO's camera eyes? What is it thinking? Nothing, of course. It is not really thinking, only processing. It doesn't really recognize old friends. It simply scans people's features and compares them to data stored in its computer. Its interactions with humans are charming and fun, but we might wish that ASIMO could smile and laugh. We might wish it looked and acted more human. Or do we?

chapter 4

Thinking Robots

Bertram, your robo-butler, rolls into the living room and announces in a flat, mechanical voice, "Dinner is served." You're slouched down in a corner of the couch. "I'm not hungry, thanks," you answer. Your parents or friends might ask if you feel all right, or if there's anything they can get you. But Bertram has no reaction. He simply rolls back into the kitchen without a word and puts away the uneaten dinner. Bertram has understood your negative reply, but he can't respond to your tone of voice or your body language. And most people, Allison Bruce discovered, really don't like that about robots.

An experimental domestic robot from the 1970s works in the kitchen.

Bruce is a researcher at the Robotics Institute at Carnegie Mellon University. Bruce is part of the institute's Social Robot program. The program studies ways to improve human-robot interaction. In the experiment, a laptop computer was attached to a robot. The robot stood in a corridor of a college classroom building and asked passing students a question. Sometimes the laptop screen would be blank, but sometimes it displayed a computer-generated face with a range of expressions.

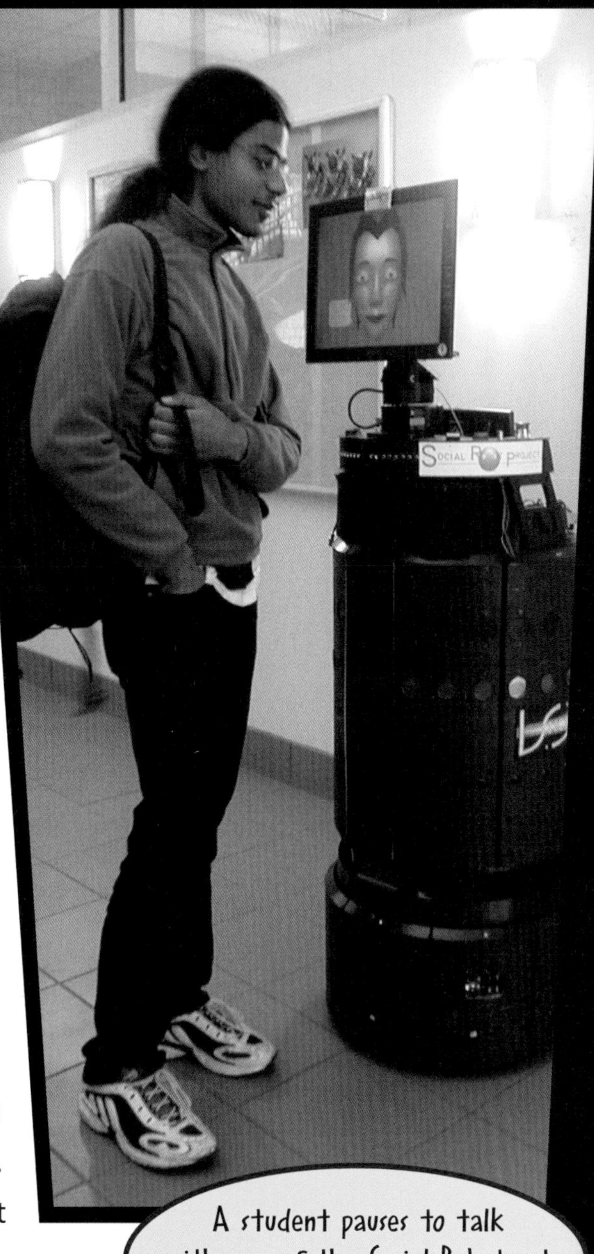

A student pauses to talk with one of the Social Robots at Carnegie Mellon University.

Bruce was not really interested in the students' answers to the questions. What she was interested in was the students' willingness to stop and talk to the robot. She found—not surprisingly—that more students responded to the robot when it had a face.

Like Bruce, other roboticists have realized that humans prefer robots they can relate to. Roboticists have developed robots that can express human emotions, such as anger, happiness, embarrassment, and sadness.

Fun with Blocks

Legos, the little plastic building bricks, are the stuff of which robots are made. One of the coolest examples of a Lego-bot is Feelix, built by Dolores Cañamero and Jakob Fredslund at the Lego Lab at Aarhus University in Denmark. Feelix is very good at expressing his feelings. His Lego eyebrows will droop and his mouth will sag when he's "sad." If he's "happy," his mouth turns into a smile, and his whole plastic brick face seems to light up.

Cañamero and Fredslund built Feelix from Lego's Mindstorms kit. The kit comes with everything needed to build a robot, including the software and instructions for programming the robot's brain through a personal computer (PC).

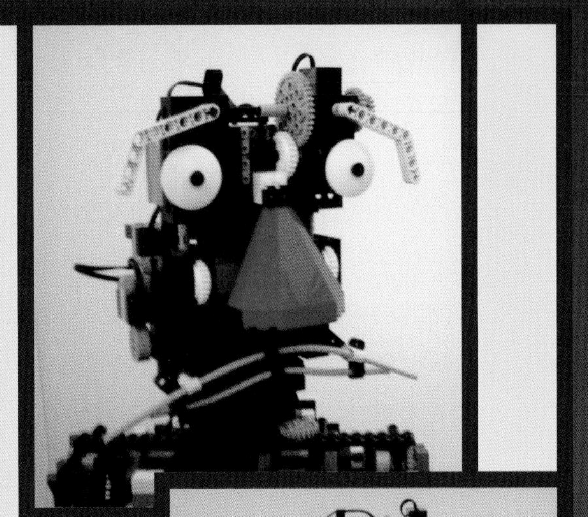

Feelix uses its eyebrows and mouth to express feelings.

I Feel, Therefore I Am

Kismet the robot was designed and built by Cynthia Breazeal, a researcher at MIT's Artificial Intelligence Laboratory. The lab is home to a whole assortment of interesting robots. But Kismet is not like the others. This robot can display emotion. Kismet's lips can pout or smile. His eyebrows can arch, and his ears can wiggle. But Kismet isn't just some puppet moving parts around. A combination of clever computer programming and sophisticated engineering has given Kismet the ability to actually respond to a stimulus in an emotionally recognizable way.

If you say words of praise to Kismet, he will smile. Bright colors also earn a smile, as does his own reflection in a mirror. But raise your voice and scold Kismet, and his lips will sink into a frown. And when Kismet becomes overstimulated by too much noise or movement, he will withdraw, lowering his eyes and taking a kind of robotic time-out.

Kismet is lovable but not just because of his blue, golf-ball-sized eyes. Kismet interacts with people and shows he has "understood" them through his facial expressions. His success in relating to people may be reflected in the fact that everyone refers to Kismet as "he" instead of "it."

Kismet was developed to interact with people.

Heads Will Roll!

His name is David Hanson, and in 2003, he showed up at a science conference in Denver, Colorado, carrying a head. The head was backless and bald and bolted to a wooden platform. But it was still rather pretty, with high cheekbones, blue eyes, and smooth rubber polymer skin. Hanson set the head down on a table, plugged it into his laptop computer, and tapped a few keys. Everyone stopped to watch what would happen.

Moments later, the head began to move, turning right and left. It smiled, sneered, and frowned. Hanson, a roboticist at the University of Texas at Dallas, called the head K-bot. To mimic the major muscles in a human face, K-bot has 24 servomotors under its specially developed rubber polymer skin. The digital cameras in its eyes watch the people who are curiously studying it, and software helps it to imitate what it sees.

David Hanson (*below, right and center*) designed K-bot to express human emotions.

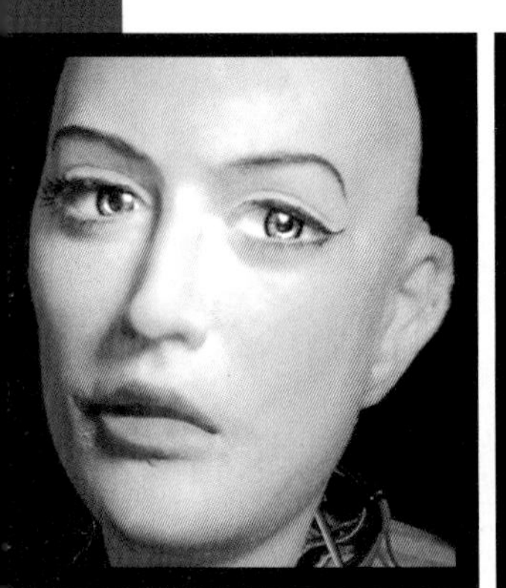

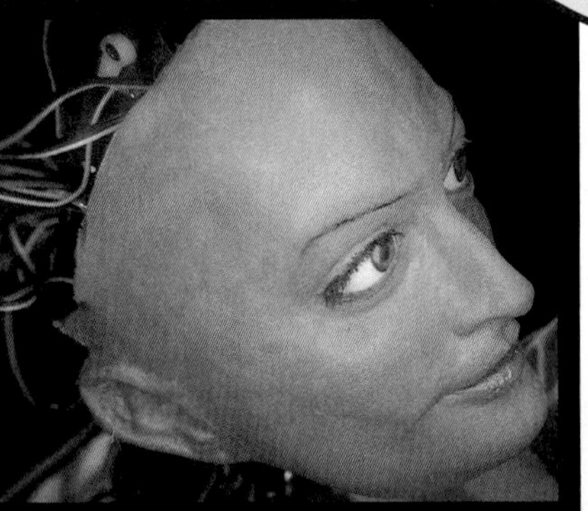

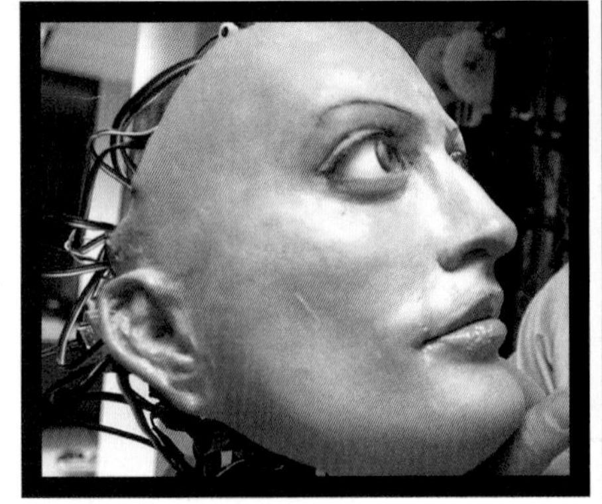

Hanson has built several robotic heads, but he isn't the only one. In Tokyo, Hiroshi Kobayashi's face robots, as he calls them, are also eerily lifelike. So is the head sitting in Fumio Hara's robotics lab at the Science University of Tokyo. Hara's robotic head can scan the face of the person standing in front of it and compare the face to those in its memory bank. Once the robot identifies which of six emotions the person is expressing, tiny machines under the robot's silicone skin remold its face to mimic what it sees.

For Hara, heads are just the beginning. His goal is to design a humanoid robot that is interactive, friendly, and, most of all, familiar. But do we really want a robot that looks just like us? Maybe not.

Fumio Hara poses with a skeleton of one of his face robots (*above*). The face robot's network of wires and pulleys are covered in a flexible skin (*below*).

In the late 1970s, Japanese robot engineer Masahiro Mori did some fascinating research on how human beings interact with robots. Mori discovered that people like friendly looking mechanical robots. But Mori found that when robots look too humanlike, people stop liking them. Mori called this sudden shift the Uncanny Valley, the place where people begin to feel uncomfortable with humanlike robots.

Mind and Matter

As scientists try to "teach" robots emotional understanding, they are also trying to program robots to think for themselves. It's called AI, artificial intelligence, and it's a tall order, indeed. AI is the attempt to get a computer to work like the human brain. If we can do that, we will be able to build the ultimate robot—an artificial version of ourselves.

As early as 1944, Alan Turing spoke of his plans to "build a brain."

In 1950, British scientist Alan Turing said that there were two ways to go about creating artificial intelligence. In the top-down approach, human beings teach the computer everything they know. The other method, called bottom-up, mimics the way we learn—a little at a time, building upon previous knowledge and experiences. You might imagine top-down as swallowing an entire library and bottom-up as snacking on one book at a time, starting with a first-grade reader.

In the 1950s, most scientists, including Turing, preferred the top-down approach. Just write a program that tells a computer what to do, step-by-step. And if you plopped that computer into a robot, there would never be any surprises because the programmer would always know what the robot's computer brain knew.

In 1966, scientists at the Stanford Research Institute (renamed SRI Technology in 1970) in California began building Shakey. This was the first intelligent, autonomous (operating competely on its own) robot. Shakey was actually able to plan the steps needed to perform simple tasks, such as moving wooden blocks around. The problem was, Shakey took forever to do anything. Its computer was so slow that it spent most of its time thinking.

Since then, computers have gotten very speedy indeed. But they are still incredibly slow compared to the processing power of the human brain. And computers don't work like a human brain. Humans learn from experience and store that information for the future. They can also quickly adapt their thinking to different situations. So if we want our robots to think the way we do, we're going to have to build a computer that can learn the way we do. And this, believe it or not, is slowly but surely happening.

Experience Is the Best Teacher

Engineers have begun building robots that can adapt to their environment. They operate on what are called patterns of behavior.

Most of these robots are quite small and behave a lot like insects. Insects don't really think. They rely on their senses and instincts to find food and to survive in an ever-changing world. Like insects, the little

A robotic ladybug, developed by Sanyo Electric Company, sits on a leaf.

insect-bots have been equipped for sensing their physical environment. But they have not been preprogrammed with any data about their environment. So when they are first turned on, they're brainless.

But the insect-bots' computers are programmed with separate "layers of behavior." The behavior layers help an insect-bot learn about its environment. The more it learns, the more it can do. Once the insect-bot has mastered one layer of behavior, the next higher layer of behavior kicks in. With each layer, the insect-bot gets better at dealing with the world around it.

Flying Robots

Since 1998, engineers at the University of California, Berkeley, have been working on a micromechanical flying insect (MFI) project. To design their MFIs, the engineers study how real flying insects lift off, use their wings and bodies, and move through the air. The engineers continue to work on making the MFIs autonomous, lightweight, and flexible.

At MIT, James McLurkin has built robot ants using these layers of behavior. But McLurkin's ant-bots are even more amazing because they are able to signal each other when they find ant-bot "food." In other words, the ant-bots learn how to cooperate to achieve a shared goal. The ultimate ant-bot, however, is yet to come—one that can communicate with real ants.

"I believe," said Hans Moravec, a research professor at Carnegie Mellon, "that robots with human intelligence will be common within the next 50 years." Certainly, the Center for Intelligent Systems (CIS) at Vanderbilt University in Tennessee shows how close we are getting. The CIS has developed a robot called ISAC (for Intelligent Soft Arm Control). ISAC can express emotion and has both short-term and long-term memory. And because this robot's brain has been designed to "think" much like ours, ISAC may soon actually be able to dream.

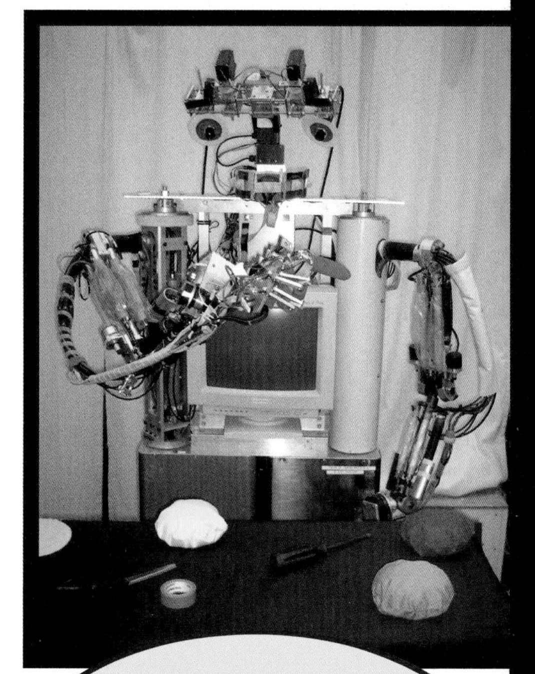

ISAC (*above*) was built to help people who have physical handicaps. Improvements focus on learning skills and response to human emotions.

It seems almost certain that in the future we will be sharing our planet with a strange new form of robotic life. What we build in the lab will have the potential to become as smart as we are. It may even improve upon its own technology. Will we love these robots or fear them? Time will tell.

Glossary

android: a mobile robot built in human form

artificial intelligence: the study of how computers can be designed to work and learn like a human brain, for use in robotics

computer: an electronic machine designed to store and process information using a mathematical system

humanoid: having human characteristics, such as facial features

Internet: an information system of interconnected computers. On the Internet, a computer user can communicate with other Internet users and access information from thousands of resources.

manipulators: the type of robot most commonly used in factories. Manipulators are usually simple robots that perform one type of job, such as welding cars or packing boxes.

medical robots: robots designed to help doctors in diagnosing and treating patients

patterns of behavior: an approach to artificial intelligence in which computers are programmed to learn information by experience

robotics: the science of designing and creating robots

robots: machines that can perform tasks and jobs through computer programming and movable parts

service robots: robots designed to interact with humans and perform helpful tasks in homes or businesses

survival instinct: a function programmed into some computers that allows them to act to avoid damage or destruction

Uncanny Valley: the point at which humans become uncomfortable with the lifelike qualities of a humanoid robot

Source Notes

For quoted material:

p. 40 (caption), The Alan Turing Home Page, http://www.turing.org.uk/turing (October 2004).

p. 43, Hans Moravec, *Mind Children,* Cambridge, MA: Harvard University Press, 1988.

Selected Bibliography

Aylett, Ruth. *Robots: Bringing Intelligent Machines to Life.* Hauppauge, NY: Barron's Educational Series, 2002.

Bruce, Allison, Illah Nourbakhsh, and Reid Simmons. *"The Role of Expressiveness and Attention in Human-Robot Interaction."* Proceedings of the IEEE International Conference on Robotics and Automation (ICRA' 02), May 2002.

Kirsner, Scott. *"They're Robots? Those Beasts!" New York Times,* September 16, 2004, E1.

Mars Exploration Rover Mission, Jet Propulsion Laboratory of the National Aeronautics and Space Agency, http://marsrovers.jpl.nasa.gov (October 2004).

Moravec, Hans. *Mind Children.* Cambridge, MA: Harvard University Press, 1988.

Paul, Gregory S., and Earl D. Cox. *Beyond Humanity: CyberEvolution and Future Minds.* Rockland, MA: Charles River Media, Inc., 1996.

Quinlan, Heather. "Oh, the Humanity!" The Learning Channel Online, http://tlc.discovery.com/convergence/robosapiens/article.html (May 2004).

Telotte, J. P. *Replications: A Robotic History of the Science Fiction Film.* Chicago: University of Illinois Press, 1995.

Whitehouse, David. *"Tiny Robots Flex Their Plastic Muscles,"* British Broadcasting Corporation (BBC) News Online, http://news.bbc.co.uk/1/hi/sci/tech/811314.stm (June 29, 2000).

Further Reading and Websites

Breazeal, Cynthia L. *Designing Sociable Robots.* Cambridge, MA: The MIT Press, 2004.
This book details Breazeal's efforts to create socially intelligent robots. The accompanying CD-ROM focuses on Kismet.

Kismet

http://www.ai.mit.edu/projects/humanoid-robotics-group/kismet/kismet.html
The Humanoid Robotics Lab at MIT runs a website dedicated to Kismet, the "sociable robot." The site includes photos and videos, information about the project, and short biographies of the people involved in Kismet's design.

The Leg Laboratory at MIT

http://www.ai.mit.edu/projects/leglab/
The Leg Lab's website features photos of individual robots, animated demonstrations of robot movement, and information on the teachers and students who work at the lab.

Mars Exploration Rover Mission

http://marsrovers.jpl.nasa.gov/home/index.html
NASA's Jet Propulsion Laboratory website features the latest news on Spirit and Opportunity. Photos from the surface of Mars, information about the planet, and details of the Rover project are included.

Robot Hall of Fame

http://www.robothalloffame.org/
Started by Carnegie Mellon University, this website honors landmark creations in robotics. The two hall of fame categories are Robots from Science and Robots from Science Fiction.

Robotics

http://www.thetech.org/robotics/
The Tech Museum of Innovation's Robotics website features a robotics timeline, a history of many types of robots, and an essay on the ethical issues behind robotic technology.

Index

Photo Acknowledgments

Photographs are used with the permission of: © Forest J. Ackerman Collection/CORBIS, p. 4; © John Springer Collection/CORBIS, p. 5 (top); © 20th Century Fox/ZUMA/Corbis, p. 5 (bottom); © Bettmann/CORBIS, p. 7; Musée d'art et d'histoire, Neuchâtel (Suisse), p. 8; University of Pennsylvania's School of Engineering and Applied Science, p. 9; © Charles O'Rear/CORBIS, pp. 10, 11; The Robotics Institute, Carnegie Mellon University, p. 12; © Getty Images, pp. 13, 22 (top), 23, 34; © Peter Russell; The Military Picture Library/CORBIS, p. 15; NASA, p. 16; © ANDY CLARK/Reuters/Corbis, p. 17; © Handout/Reuters/Corbis, p. 18 (top); © NASA/JPL/Handout/ Reuters/Corbis, p. 18 (center); © AFP/Getty Images, pp. 18 (bottom), 22 (bottom); © Reuters/ Corbis, pp. 19, 25; © Royalty-Free/CORBIS, p. 20; © AP Wide World Photos, pp. 21, 38 (top center and right), 38 (bottom left and center); courtesy of Daniel Kingsley, Roger Quinn, and Roy Ritzmann, Case Western Reserve University, p. 26 (top and center); © Tom Young/CORBIS, p. 26 (bottom); Donna Coveney/MIT, p. 28; courtesy of American Honda Motor Co., Inc., ASIMO division, p. 30; © Ramin Talaie/Corbis, p. 31 (top); © Haruyoshi Yamaguchi/Corbis, p. 31 (bottom); courtesy of Tucker Balch, GVU Center, Georgia Institute of Technology, p. 32; courtesy of Sony Electronics, Inc., p. 33; courtesy of Alison Bruce, The Robotics Institute, Carnegie Mellon University, p. 35; © Jakob Fredslund, p. 36; © Rick Friedman/Corbis, p. 37; © David Hanson, p. 38 (top left and bottom right); © Peter Menzel Photography, p. 39; courtesy of Lady Nicola Turing, p. 40; © MASTER PHOTO SYNDICATION/CORBIS SYGMA, p. 42; courtesy of Flo Fottrell and Kazuhiko Kawamura, Center for Intelligent Systems, Vanderbilt University, p. 43.

Cover: © AP Wide World Photos (main photo); American Honda Motor Co., Inc., ASIMO division (top); NASA (center); courtesy of Flo Fottrell and Kazuhiko Kawamura, Center for Intelligent Systems, Vanderbilt University (bottom).